Bibliographic information published by the German National Library:

The German National Library lists this publication in the National Bibliography; detailed bibliographic data are available on the Internet at http://dnb.dnb.de .

Imprint:

Copyright © 2004 GRIN Verlag, Open Publishing GmbH
Print and binding: Books on Demand GmbH, Norderstedt Germany
ISBN: 978-3-656-81489-4

This book at GRIN:

http://www.grin.com/en/e-book/57824/air-transportation-deregulation

Kimberly Wylie

Air Transportation Deregulation

GRIN Publishing

GRIN - Your knowledge has value

Since its foundation in 1998, GRIN has specialized in publishing academic texts by students, college teachers and other academics as e-book and printed book. The website www.grin.com is an ideal platform for presenting term papers, final papers, scientific essays, dissertations and specialist books.

Visit us on the internet:

http://www.grin.com/

http://www.facebook.com/grincom

http://www.twitter.com/grin_com

Air Transportation Deregulation

By: Kimberly Wylie

University of Phoenix

Table of Contents

Abstract 1

Introduction: 2

Deregulation: 2

Civil Aeronautics Board: 3

Struggling Through Deregulation: 4

Conclusion: 5

References: 6

Abstract

One of the most significant events in the air transportation industry was the Airline Deregulation Act of 1978. This paper will review what deregulation, in general is, and then move specifically to the Airline Deregulation Act of 1978. From there, this paper will explore what effects the Act has had on the industry. By understanding these concepts, one can better understand the environment the industry now operates in, and how far it has come.

Air Transportation Deregulation

Introduction:

One of the most significant events in the air transportation industry was the Airline Deregulation Act of 1978. This paper will review what deregulation, in general is, and then move specifically to the Airline Deregulation Act of 1978. From there, this paper will explore what effects the Act has had on the industry. By understanding these concepts, one can better understand the environment the industry now operates in, and how far it has come.

Deregulation:

Deregulation is, in its most simplistic terms, when the government removes certain regulations on businesses, to encourage the industry. The premise behind deregulation is that with fewer regulations impeding operations, businesses will become more competitive. This increased competition will then result in higher productivity levels, increased efficiencies, and lower prices for consumers ("Deregulation", 2004).

There have been some failures in attempts to deregulate certain industries. Most notably, the deregulation of the Savings and Loan industry in the 1980s was perceived as a failure and was subjected to re-regulation in order to balance out the industry. The Savings and Loan scenario is a perfect example of why regulation or deregulation must be handled responsibly and intelligently, taking into account and utilizing sophisticated economic theories to help predict the future results ("Deregulation", 2004).

The Airline Deregulation Act was signed into law on October 28, 1978. The Act was created to remove governmental control and open up the passenger air transport industry to free market forces. As noted above, the desire was to increased efficiencies within the industry by promoting competition, which would hopefully lead to reduced airfares for consumers ("Airline Deregulation", 2004).

Civil Aeronautics Board:

Before the Act was enabled, the Civil Aeronautics Board (CAB) regulated all domestic air travel. They were in control of establishing fares, setting air traffic routes, and creating airline schedules. "The CAB promoted air travel, and due to their beliefs they held fares down in the short-haul market and let them be higher in the long-haul market. The CAB was also obliged to ensure that the airlines had a reasonable rate of return" ("Airline Deregulation", 2004).

However, Congress, as well as others, began to see that the CAB was inhibiting the growth of the industry. They saw that because of the regulations that were in place, the CAB was actually encouraging inefficiencies. This was exacerbated in the 1970s when skyrocketing oil prices and inflation led to increased fares and capacity moratoriums. Congress felt that by deregulating the industry, the market forces within the industry would determine pricing, and improve the quantity and quality of air services within the United States. They also saw the potential for these market forces to reduce the inflated long-haul fares, as well as lower the barriers to entry for new airlines entering the industry ("Airline Deregulation", 2004).

The deregulation took place faster than the anticipated four years schedule for complete deregulation of domestic routes. The CAB passed on many of its functions to the Department of Transportation (DOT), including the ability to grant antitrust immunity. In addition they were charged with the duty of ensuring smaller communities would still be serviced by air carriers, via the Essential Air Service program, by offering subsidies ("Airline Deregulation", 2004).

As anticipated, the consumer did benefit from the deregulation. A General Accounting Office report conducted in 1996, reported that the average fare per passenger mile was actually 9 percent lower in 1994 than in 1979. However, airlines were struggling to survive in their newly created free market ("Airline Deregulation", 2004).

Struggling Through Deregulation:

By 1996, the number of airline carriers had nearly doubled creating a hyper-competitive environment. New, smaller airline entered the market rapidly, driving down prices ("The airline bankruptcies", n.d.) This increased competition caused nine major air carriers to go bankrupt between the years 1978 and 2001. These included: Eastern, Midway, Braniff, Pan Am, Continental, America West, and TWA. To add to the financial pandemonium, more than 100 smaller airlines went bankrupt as well ("Airline Deregulation", 2004).

In order to capitalize on economies of scale, mergers and acquisitions began to take place, creating several 'mega-carriers' and oligopolistic conditions. To increase efficiencies, many airlines converted their services to hub-and-spoke systems, giving up much of their point-to-point systems of regulation days ("Airline Deregulation", 2004). Many smaller cities were abandoned in the process of adopting hub cities ("The Airline bankruptcies", n.d.)

Although deregulation caused increased competition in the industry, it is a bit unfair to blame deregulation on such mismanagement blunders as were the cases of Eastern, Continental and Pan Am. In the case of Eastern Airlines and Continental, deregulation was less to blame then one man, Frank Lorenzo. Lorenzo, recognizing that Continental was struggling due to the increased competition after deregulation, took over the airline under his holding company, Texas Air Corporation. Merging his airline, Texas International Airlines, with Continental, Lorenzo began to slash costs haphazardly. He filed for Chapter 11, allowing him to throw out any previous labor union agreements employees had, cutting salaries by as much as 50 percent, and reducing benefits for key employees such as pilots and executives. The mid-1980s saw Lorenzo profiting from this price-slashing method; however, discontent was growing in the ranks ("The airline bankruptcies", n.d.)

Lorenzo's takeover of Eastern Airlines, in 1986, was short-lived. The company was already in significant debt due to mismanagement and labor unrest. Lorenzo's actions did not help the situation. Due to Lorenzo's over expansion, blossoming debt, and labor conflicts, by 1990, Lorenzo had quit the airline, and left it to falter and die a year later, with $3.2 billion in debt. Clearly, deregulation did not help either carrier's situation, but it was the poor management decisions in over expansion and poor labor relations that eventually cost the company.

Pan Am is another case that many think of when discussing the ill-effects of deregulation. Pan Am's narrow focus of international routes was their biggest mistake. With new competition in this area of service, they faced intensified price wars.

Recognizing their mistake, Pan Am acquired National Airlines, in order to service domestic routes. However, increased fuel costs, labor conflicts, and an economic recession compounded their debt problems. Despite selling off much of its assets, including the Pan Am building in New York City, and its most profitable routes, Pan Am had net operating losses totaling $3 billion by 1989. Again, although the increased competition from deregulation had been an added thorn in Pan Am's side, it was their mismanagement and poor strategic decisions that led to their ultimate demise.

Conclusion:

Now, 26 years later, the industry has stabilized as much as any industry ever does. Those that survived the initiation period of the 1980s have learned from their and their competitor's mistakes. With the increased competition from a free market, airline must constantly be vigilant to providing the most efficient and effective service to their customers. Competitive advantages are few and far between, and when they are to be had, more than likely a competitor will follow suit. Yet, deregulation has created a much more efficient industry, making air travel more affordable for everyone.

References:

The Airline bankruptcies of the 1980s. (No date). Retrieved July 28, 2004, from http://www.1903to2003.gov/essay/Commercial_Aviation/Bankruptcy/Tran9.htm.

Airline Deregulation Act. (2004, July 11). Retrieved July 28, 2004, from http://en.wikipedia.org/wiki/Airline_Deregulation_Act.

Deregulation. (2004, June 18). Retrieved July 28, 2004, from http://en.wikipedia.org/wiki/Deregulation.